Who Am I?

Mountain Animals

Written by Read With You
Center for Language Research and Development

The characters and events portrayed in this book are fictitious. Any similarity to real persons, living or dead, is coincidental and not intended by the author.

Published by Read With You Publishing

Designed by Read With You Center for Language Research and Development

ISBN-13: 978-1-944710-67-5
ISBN-10: 1-944710-67-1

Third Edition July 2019
Printed in the United States of America.

I'm a relative of the camel, but I don't have a hump.

Who am I?

llama

Only the male grows huge antlers.

Who am I?

moose

I have large, curled horns.

Who am I?

bighorn sheep

I sleep in a cave for most of the winter.

Who am I?

brown bear

I'm also known as a mountain lion.

Who am I?

puma

I have really good eyesight.

Who am I?

eagle

I can climb trees.

Who am I?

black bear

I love eating bamboo.

Who am I?

panda bear

Llama

Moose

Bighorn Sheep

Brown Bear

I am a llama.

I am a moose.

I am a bighorn sheep

I am a brown bear.

Puma

Eagle

Black Bear

Panda Bear

I am a puma.

I am an eagle.

I am a black bear.

I am a panda bear.

We are mountain animals.

The mountains are high, rocky places. The highest points of mountains are called peaks.

Made in United States
Orlando, FL
07 November 2026

90337536R00017